INVESTIGATING
Body Systems

Investigating the RESPIRATORY SYSTEM

Natalie Hyde

A CRABTREE FOREST BOOK

Crabtree Publishing
crabtreebooks.com

Author: Natalie Hyde

Series research and development: Natalie Hyde

Editorial director: Kathy Middleton

Editor: Ellen Rodger

Proofreader: Melissa Boyce

Design: Tammy McGarr

Print and production coordinator: Katherine Berti

IMAGE CREDITS

Shutterstock: Larry St. Pierre p 5; Neveshkin Nikolay p 10 (top); EQRoy p 10 (bottom); criben p 22; Amy Kerkemeyer p 27 (bottom); Alexandros Michailidis p 29 (top); Paolo Bona p 35 (top);

Wikimedia Commons: Welcome Images p 13 (bottom);

Crabtree Publishing

crabtreebooks.com 800-387-7650

 In Canada: We acknowledge the financial support of the Government of Canada through the Canada Book Fund for our publishing activities.

Hardcover 978-1-0398-0650-4
Paperback 978-1-0398-0676-4
Ebook (pdf) 978-1-0398-0702-0
Epub 978-1-0398-0729-7

Published in Canada
Crabtree Publishing
616 Welland Avenue
St. Catharines, Ontario
L2M 5V6

Published in the United States
Crabtree Publishing
347 Fifth Avenue
Suite 1402-145
New York, NY 10016

Library and Archives Canada Cataloguing in Publication
Available at the Library and Archives Canada

Library of Congress Cataloging-in-Publication Data
Available at the Library of Congress

Printed in the U.S.A./012023/CG20220815

Contents

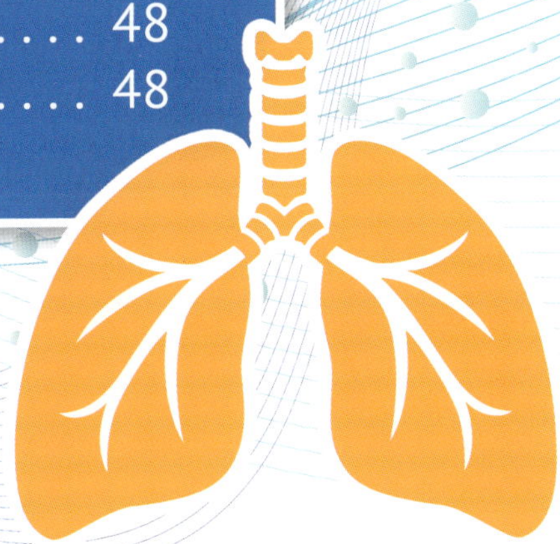

INTRODUCTION

The Respiratory System at Work

Your soccer game is in full swing. You are sprinting across the pitch to take a pass from a teammate. When you head to the bench you are out of breath, sweating, and your heart is pounding. These are all signs of your respiratory system at work. Breathing and respiration are the two processes that make up this system. Breathing takes oxygen into your body to transfer it into the blood. Respiration happens in the cells where oxygen, or the gas we breathe, is exchanged or made into another gas called carbon dioxide.

Every living thing needs to breathe: plants, **microbes**, and animals, including humans. Insects breathe through holes in their bodies. Fish take in oxygen from water through their gills. Mammals, adult **amphibians**, birds, and reptiles use lungs. Animals take in oxygen and give off carbon dioxide. Plants exchange gases the opposite way to animals. They take in carbon dioxide and give off oxygen. That is why it is vital to keep a balance between plants and animals, and forests and cities.

For everything we do in life, every move we make, even in sleep, we have to breathe. By the time we are 70, we will have taken about 600 million breaths. Our lungs are the organ we use to breathe, but they are only part of the system that allows our body to use oxygen to live, move, and grow. Breathing in is called inhaling. Breathing out is called exhaling.

How Much Do We Breathe?

The average person takes around 17,000 breaths per day. The air we inhale is about 21 percent oxygen and 78 percent nitrogen. Our bodies do not use nitrogen, so we just breathe it back out again. We also exhale about 2 cups (0.5 L) of water each day. That is why you should make sure you drink enough water daily to replace it.

NFA
17
SHS
BEARS
14

Inhale and Exhale

In order for the respiratory system to work, it has to have a way to bring air into the body and a place to store it for a while. Then it has to have a way to remove the oxygen the body needs from the stored air and replace it with carbon dioxide that it doesn't need. Lastly, it has to **expel** the unnecessary gases from the body.

The lungs are the main organ of the respiratory system. They are the storage area for inhaled air. They are a little like two balloons in your chest. When you inhale, the balloons fill up. When you exhale, they **deflate**. You can feel that movement if you put your hands on your chest. You also know your respiratory system is working on cold days when you can see your breath as you exhale. That is because the water droplets in exhaled air cool down outside the body and form a mist.

Our breathing changes over our lifetime. Newborn babies breathe from 40 to 60 times per minute. This is because their lungs take up a large portion of their chest. There isn't as much room for them to store air in between breaths. As they grow, that space increases and the average breaths decrease. Adults breathe about 16 times per minute. Both adults and babies breathe more slowly when asleep. When we are afraid or worried we breathe much more quickly—for adults, up to 100 times per minute.

A baby's breathing decreases as they grow, from 40 to 60 breaths a minute as newborns to 25 to 40 a minute at six months.

Snoring

Snoring is the loud sound that sometimes happens when people sleep. Air flows past tissue in the throat, causing the tissue to vibrate with each breath. The loudest snore ever recorded was about 112 decibels. That is as loud as a jet flying right over your house!

Breathing in deeply through the nose and out through the mouth is good for our lungs.

CHAPTER 1

How Do We Know?

Healers and physicians in ancient times did not know the purpose of breathing. They also didn't know at first that breathing was connected to the lungs.

Lungs and breathing were first mentioned together in a record written by the ancient Greeks in about 384 to 322 B.C.E. At the time, people thought that the role of breathing was to cool the heart and that most of the process happened through pores in the skin.

The first person to suspect that a vital substance is taken from air during breathing and transported in the blood to the heart was Galen. He was a Greek physician working in the **Roman Empire** around 150 C.E. Galen examined the human body when he did surgeries and **autopsies**. He was the first to understand that the **diaphragm** was used to help the lungs expand and **contract**.

Understanding that the heart and lungs work together came next. Ancient texts describe how the heart pumps blood that has come in contact with air around the body. Researchers can trace this idea back to Qusta ibn Luqa, who was born around 860 C.E. in Baalbek in what is now modern Lebanon.

In the Middle Ages, from 500 to about 1500 C.E., scientists were forbidden by the Christian Church to open bodies and investigate. Because of this, our knowledge of the human body and how it worked stalled for centuries.

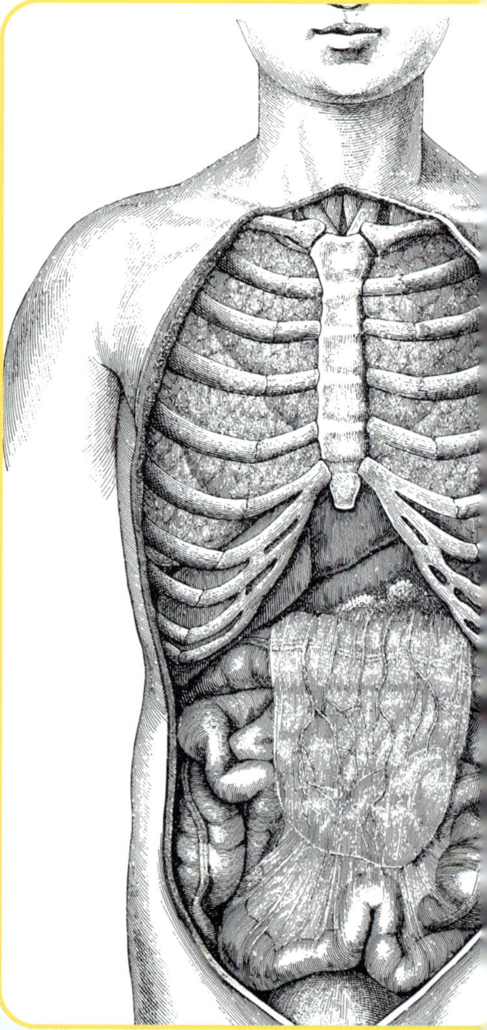

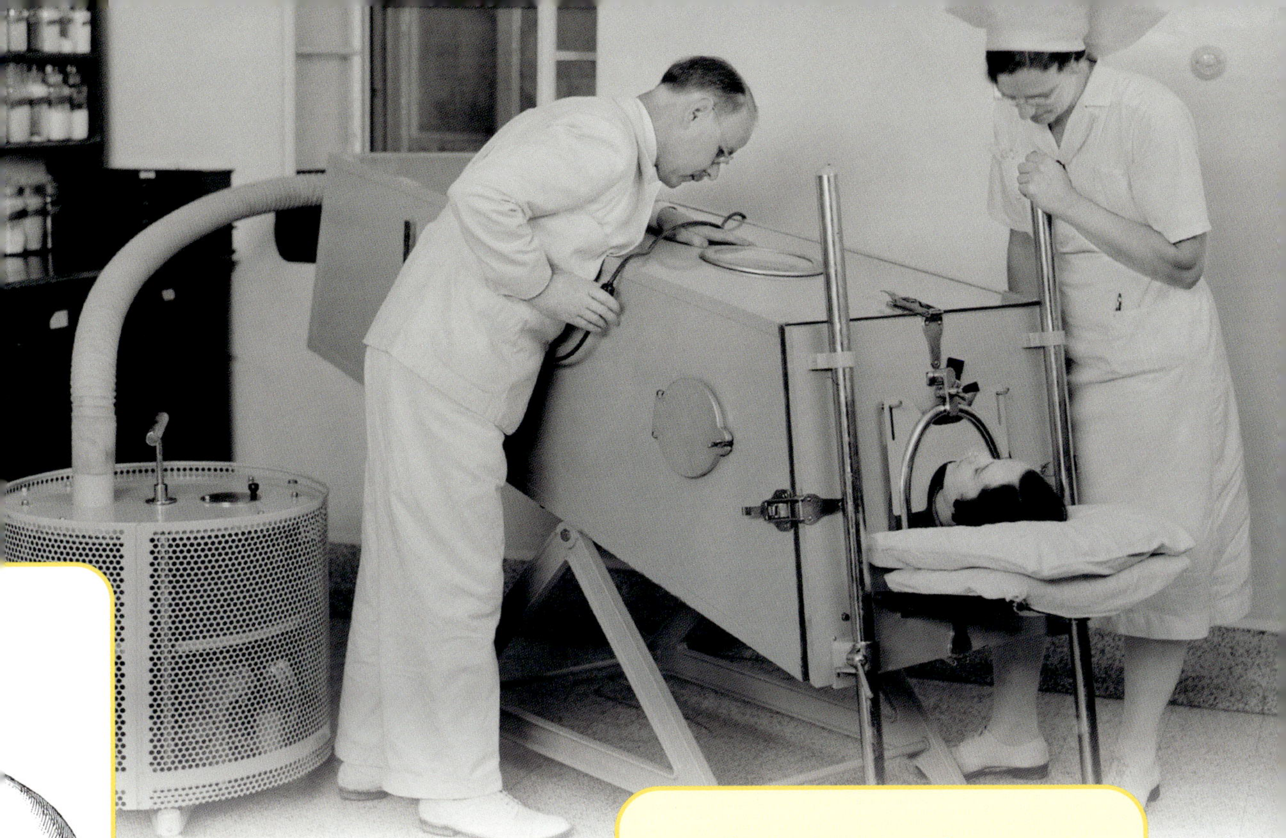

A polio patient in an iron lung at a hospital in 1940. Polio sometimes weakened the muscles used to inhale and exhale. An iron lung assisted respiration. Some patients used them for a few weeks or months. Others needed them all the time.

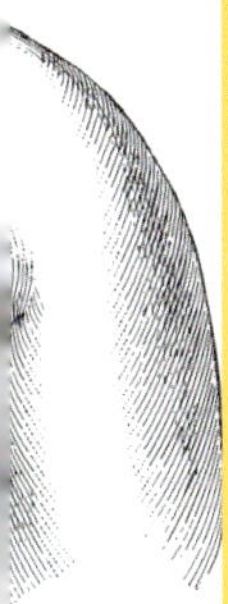

The heart and lungs are protected by the rib cage.

The Iron Lung

The iron lung is a machine that was invented by Philip Drinker and Louis Agassiz Shaw in 1928. They wanted to help children who were paralyzed from **polio** and unable to breathe. The machine is a large metal tube that surrounds a person's body, except for the head. Using pressure, it mechanically helps their chest to rise and fall, forcing air in and out, allowing them to breathe.

Blood Circulation

Our knowledge of the respiratory system started increasing again in the 1600s. In 1628, English surgeon William Harvey caused a stir when he stated that blood circulated around the body and didn't just get **absorbed** into tissue. This was a new idea and **controversial** because it suggested that the heart was just a pump and not the spot for the soul. Religion and science were at odds.

Marcello Malpighi was an Italian biologist. In 1661, while examining a frog with a microscope, he discovered the very tiny tubes in the lungs called capillaries. He could see the blood flowing in and out of them. But scientists still weren't sure what substance was being transferred to the bloodstream in the lungs.

The idea that breathing was an exchange of gases, not liquids, came in the 1770s. Antoine Lavoisier was a French chemist who discovered the gases that moved in and out of the body in the air we breathe. But he still believed that oxygen was consumed in the lungs, and did not travel farther than that. It took another 100 years for scientists to understand that the gases themselves entered the bloodstream inside the lungs.

The discovery of X-rays in 1895 allowed doctors to see under the skin for the first time. Doctors were able to "see" lung diseases such as tuberculosis. Tuberculosis is a disease caused by bacteria that can be passed from person to person. Studying the lungs of patients who died of the disease helped researchers understand how the lungs work.

WILLIAM HARVEY

MARCELLO MALPIGHI

ANTOINE LAVOISIER

Phlogiston to Oxygen

In the 1700s, scientists came to the conclusion that something vital to life in animals was present in the air but they didn't know what it was. This unknown substance was named "phlogiston." Antoine Lavoisier later renamed it oxygen.

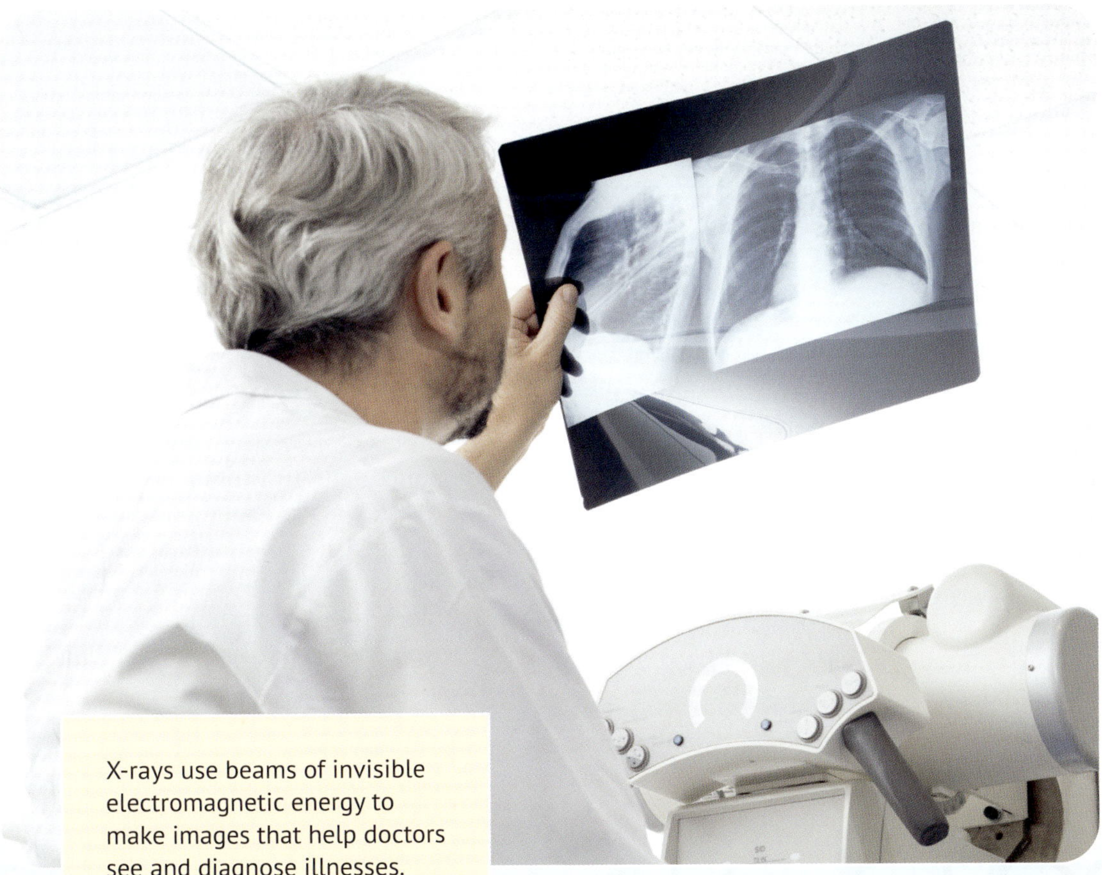

X-rays use beams of invisible electromagnetic energy to make images that help doctors see and diagnose illnesses.

CHAPTER 2

How It Works

The respiratory system consists of two parts: breathing, also called the pulmonary system, and respiration, which is the exchange of gases in and out of blood.

Breathing is a physical process. Respiration is a **chemical process**. Breathing takes place in the lungs and respiration takes place in the cells. The respiratory system's main function is to provide oxygen to our blood cells to allow them to work. But this system also has other functions:

- It allows us to talk:
  The structures that produce sound are found in the passageways that allow the air we inhale to reach our lungs.
- It allows us to smell:
  The cells that can detect odors are found in the air passageways of our nose.
- It warms the air:
  Our normal body temperature is 98.6 °Fahrenheit (37 °C). As air travels in through our respiratory passageways, it is warmed up or cooled down to match our body temperature.
- It adds moisture to the air we breathe:
  Our bodies are made mostly of water. To prevent dry air from damaging tissue, the respiratory system adds moisture to match the **humidity** level of the body.
- It removes waste gases:
  As our bodies work, oxygen is used up and carbon dioxide is produced. Nitrogen is also inhaled. These waste gases need to be exhaled from our bodies.
- It protects airways from harmful substances and **irritants**:
  Germs like viruses, bacteria, fungi, and toxins are also sometimes present in the air. Substances in the respiratory system can block these harmful irritants from entering the system and flush them out of the body.

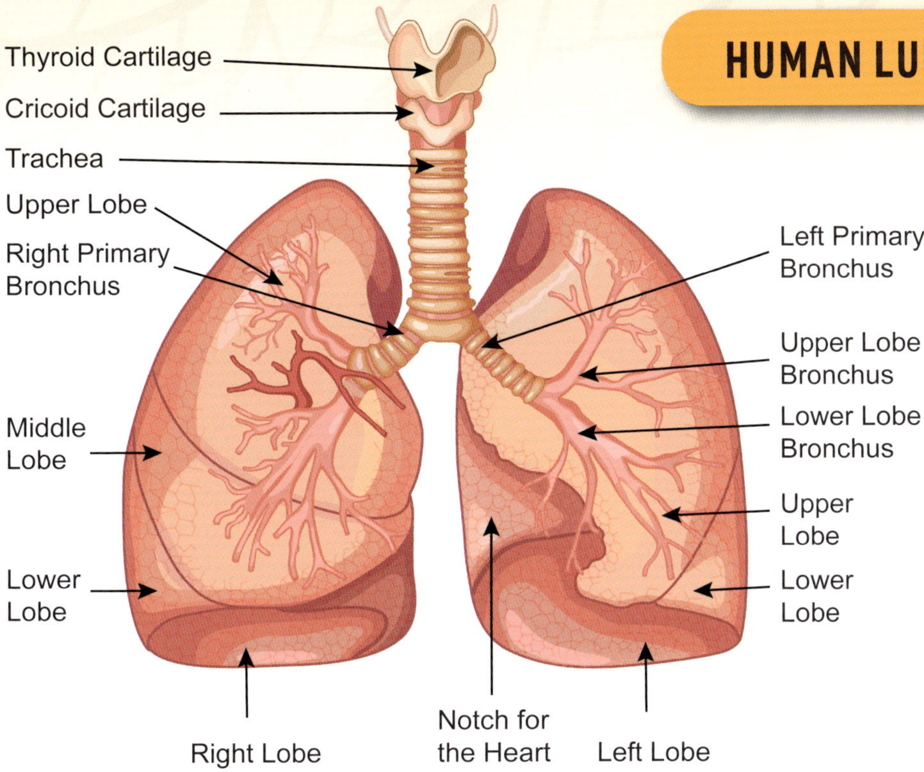

Key Figures

Rene Laennec is considered the "Father of Pulmonary Medicine." He invented the **stethoscope**, which allows doctors to listen to sounds inside bodies. Using that, he was able to accurately describe the different sounds of normal and diseased lungs.

Upper and Lower Tracts

The respiratory system is a network of organs, bones, and muscles. The structures found in the upper part of the body are called the upper respiratory tract. The organs and structures below these are called the lower respiratory tract.

The upper respiratory tract includes the nose, nasal cavity, mouth, pharynx (throat), and larynx (voice box). The upper respiratory tract provides a route for air to move from the atmosphere to the lungs. It is here that substances or organisms harmful to the body can be blocked or removed. The air is cleaned, warmed, and humidified, or made moist, before reaching the sensitive lungs. The gases present in the air, however, are unchanged.

The lower respiratory tract starts with the trachea (windpipe), and includes the lungs, diaphragm, and rib muscles. The air is pulled into the lungs where oxygen is removed and sent into the bloodstream. Then waste gases are collected and the rib muscles and diaphragm push the air back up into the upper respiratory tract where it can leave the body.

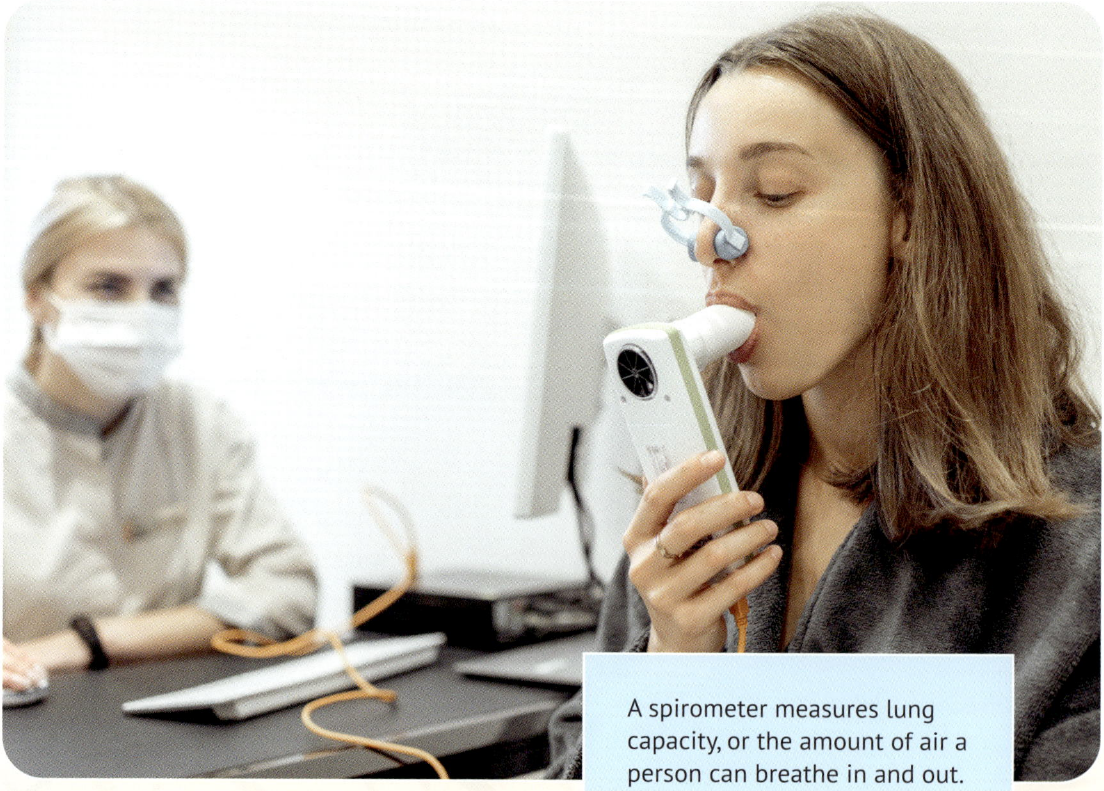

A spirometer measures lung capacity, or the amount of air a person can breathe in and out.

Sneezing

Sneezing is how the body gets rid of air that might contain things that irritate the nose and throat, such as dust, dirt, or **allergens**.

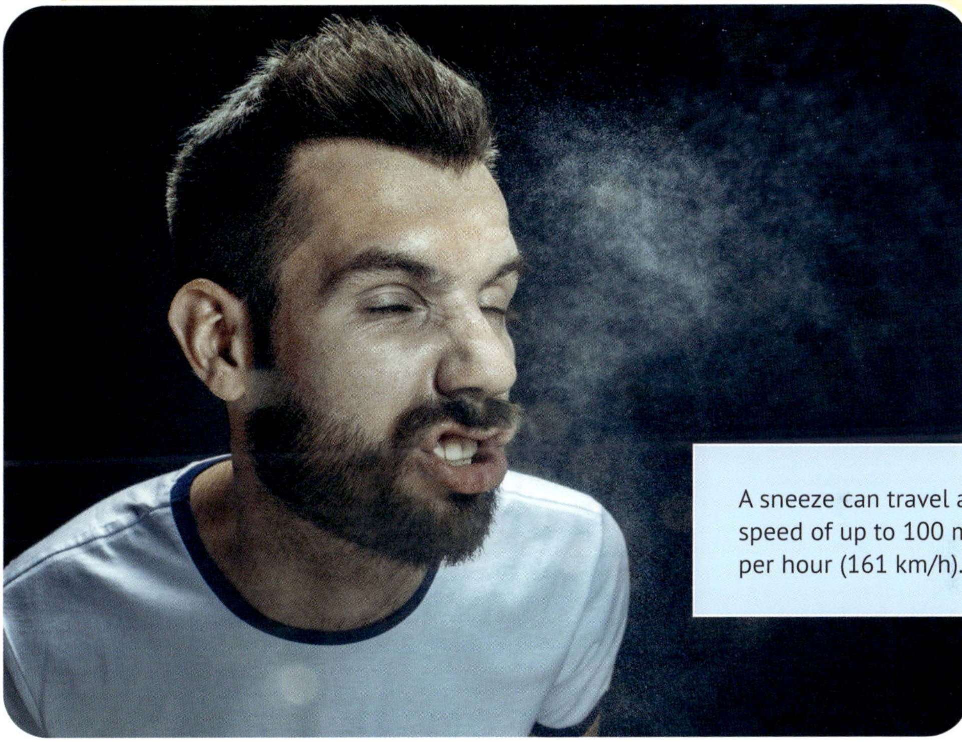

A sneeze can travel at a speed of up to 100 miles per hour (161 km/h).

How Are These Connected?

The nose and mouth meet at the pharynx at the back of the throat. At the top of the pharynx is the larynx, which is how we make sound. The pharynx divides into the **esophagus** for food (heading to the stomach) and the trachea for air (heading to the lungs.) The epiglottis is the small flap that covers the trachea so food and liquids do not accidentally get into the lungs. The lungs are protected by the bones of the rib cage. The diaphragm is the domed-shaped muscle below the lungs, and the rib muscles are between the rib cage bones, surrounding the lungs.

Look Inside the Lungs

Healthy lungs have the same texture as a squishy sponge and are a light-pink color. The lungs consist of two sacs of tissue called lobes. They are covered by two thin layers of tissue called pleura that protect the soft insides. The lobes connect to each side of the trachea so that the lungs sit on the left and right sides of the chest.

The trachea divides into two tubes that each connect to a lobe, forming an upside-down "Y." These tubes are called bronchi. The larger bronchi then divide into smaller and smaller branches called bronchioles. The bronchioles stretch out to reach every part of the lung tissue, dividing to become smaller and smaller until they can only be seen with a microscope.

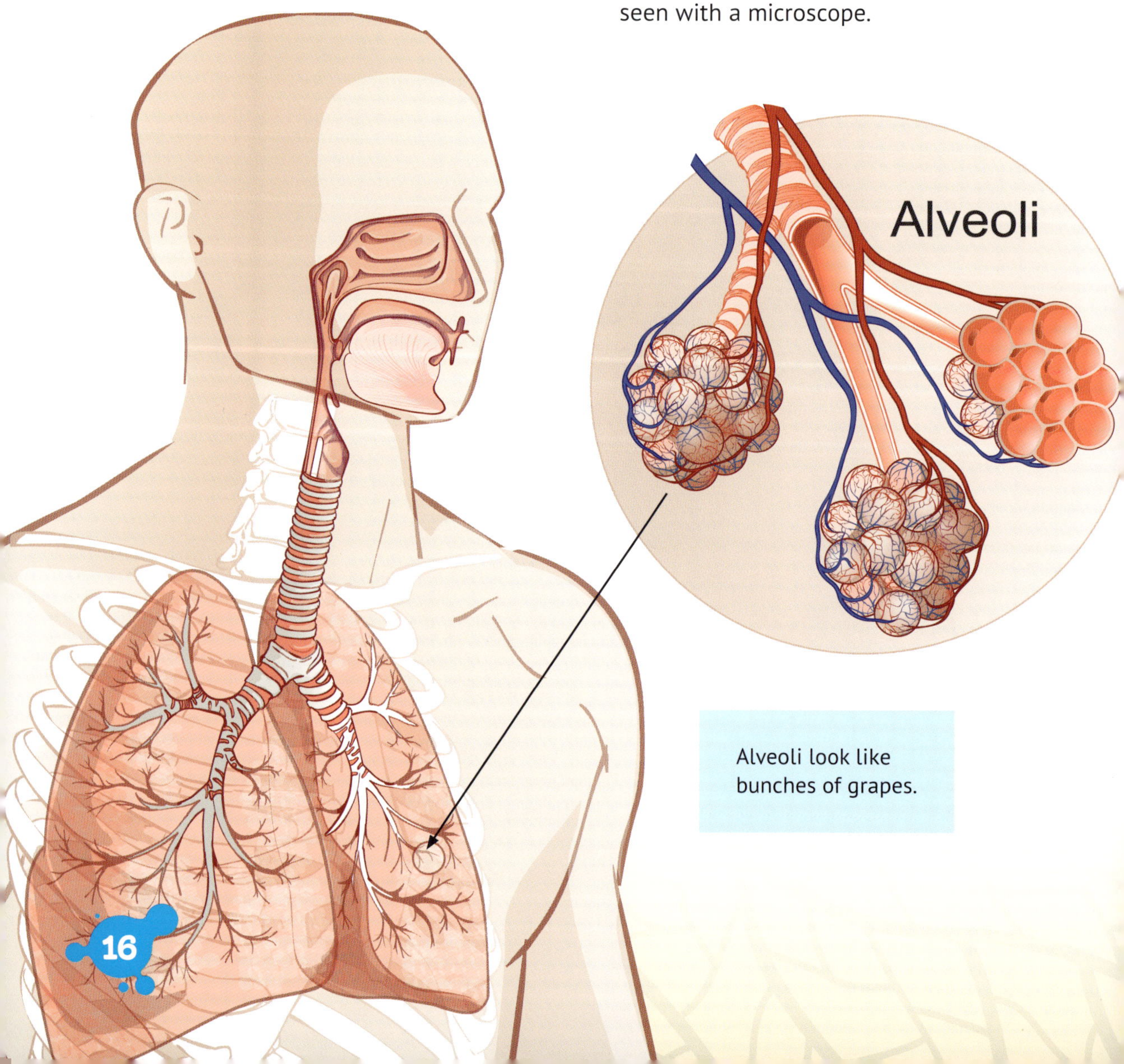

Alveoli look like bunches of grapes.

At the end of the tiniest bronchioles are a cluster of small air sacs called alveoli. They look like a bunch of tiny grapes. There are millions of alveoli in the lungs. In the walls of the alveoli are tiny blood vessels called capillaries. The oxygen in the air in the lungs moves through the walls of the alveoli and into the bloodstream. The capillaries in turn are connected outside the alveoli to larger blood vessels that move the oxygen-rich blood to the cells in the body and then bring back the waste gases.

Fact Check

Is yawning a sign that you are tired?

FACT: Researchers say that in fact yawning is a signal that the body is trying to make you more alert. Yawning releases **hormones** that briefly increase your heart rate and alertness.

I've Got Rhythm

Many different muscles and tissue have to work together in rhythm for us to breathe. All of this is controlled by a section of our brains called the medulla oblongata.

The medulla oblongata is at the base of the **brain stem**. It sends signals to the muscles that control breathing, such as the diaphragm and rib muscles. It also controls other air movement in the body such as sneezing and coughing.

In order to inhale, the brain tells the diaphragm to contract and move downward and the rib cage to rise. This allows room for the lungs to expand with air. When we exhale, the brain signals the diaphragm to relax and move upward while the rib muscles help push the air back out of the lungs. Some of the air we breathe in doesn't make it to the alveoli to exchange gas. This is called "dead air."

Inhaling can be done through the nose or mouth. The nose can produce a sticky substance called mucus. Mucus traps particles in the air that might be damaging for the body. Little hairs called cilia line the nostrils. The hairs move in rhythm to flush the mucus out of the nose. The liquid produced in the mouth, called saliva, has a similar role as mucus. It helps flush unwanted particles out of the mouth.

Breathing IN

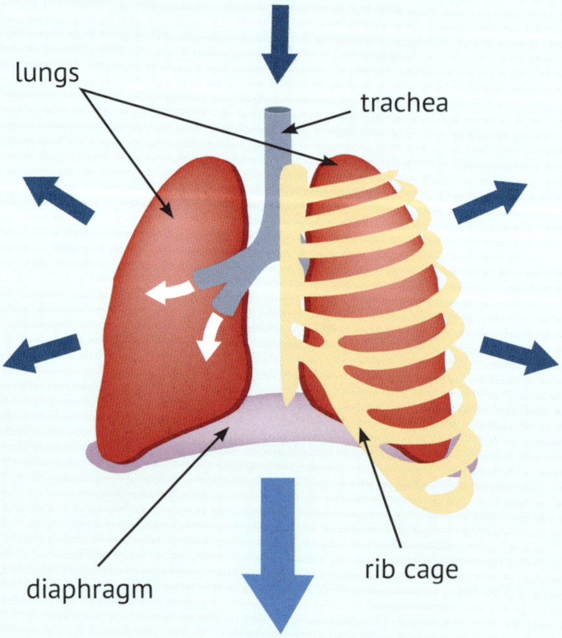

Inspiration

Breathing OUT

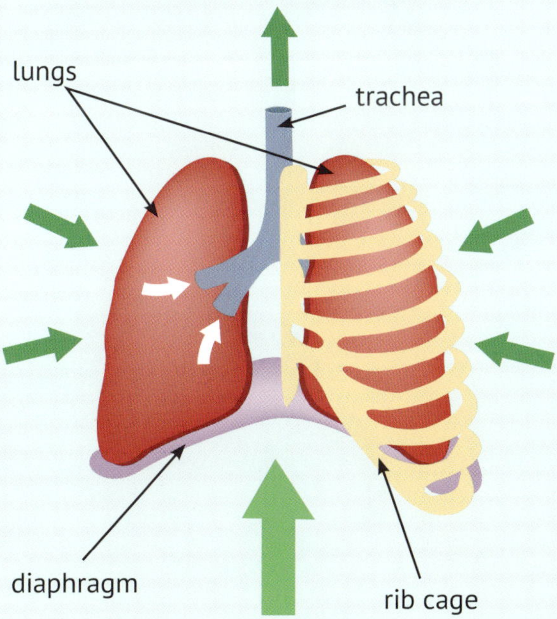

Exhalation

Fact Check

Are both our lungs the same size?

FACT: No, the left lung tends to be narrower than the right in order to make room for the heart, which is located on the left side of the body. The right lung is a little shorter in order to leave room for the liver, which is right underneath it.

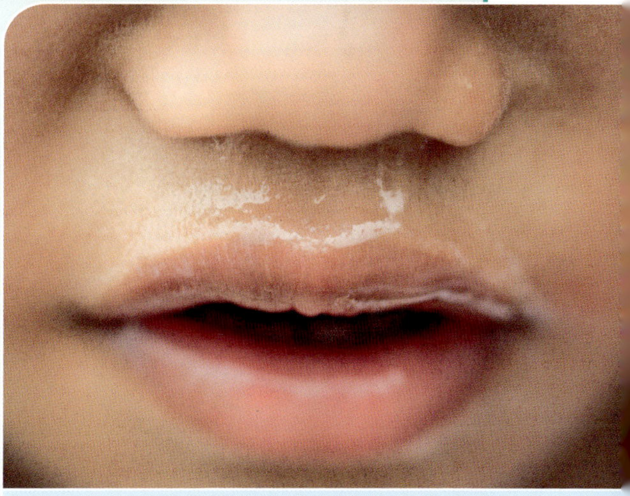

Mucus is important for flushing unwanted things out of the nose.

Gas Exchange

Respiration happens in the alveoli. The walls of the alveoli are only one cell thick. This is what allows the gas to pass through the walls and into the blood in the attached capillaries.

The moving of gas through the walls of the alveoli is called diffusion. The bloodstream then carries the oxygen around the body in red blood cells, providing energy. Carbon dioxide is carried back to the alveoli and the gas moves through the cell wall to join the air ready to be exhaled out of the body. We don't even have to think about this regular process—the medulla oblongata does it automatically. Sometimes we exhale on purpose, such as when we blow out a candle. When that happens our brain **deliberately** tells the muscles in our abdomen to contract and force air out of our lungs.

The gases in our system need to stay in balance. When we exhale too often, such as when we are frightened, we can lower the carbon dioxide in the blood. This is called hyperventilation. Often when people are hyperventilating they are given a paper bag to breathe into. By breathing in their own exhaled air, the carbon dioxide will help their body regain its balance of gases.

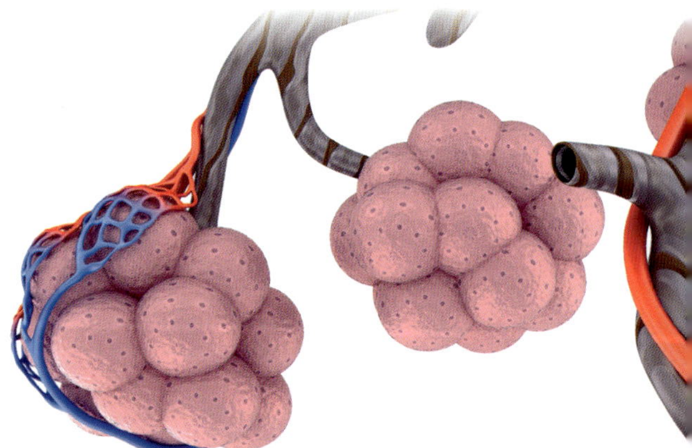

Fact Check

Does breathing out get rid of all the air in the lungs?

FACT: No. Even after exhaling, about 0.26 gallons (1.2 L) of air is still in the lungs. If all the air was out, the lungs would collapse.

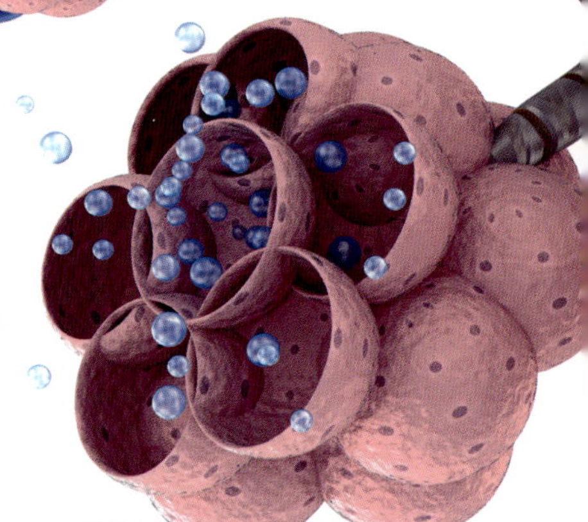

Alveoli saturating blood with oxygen

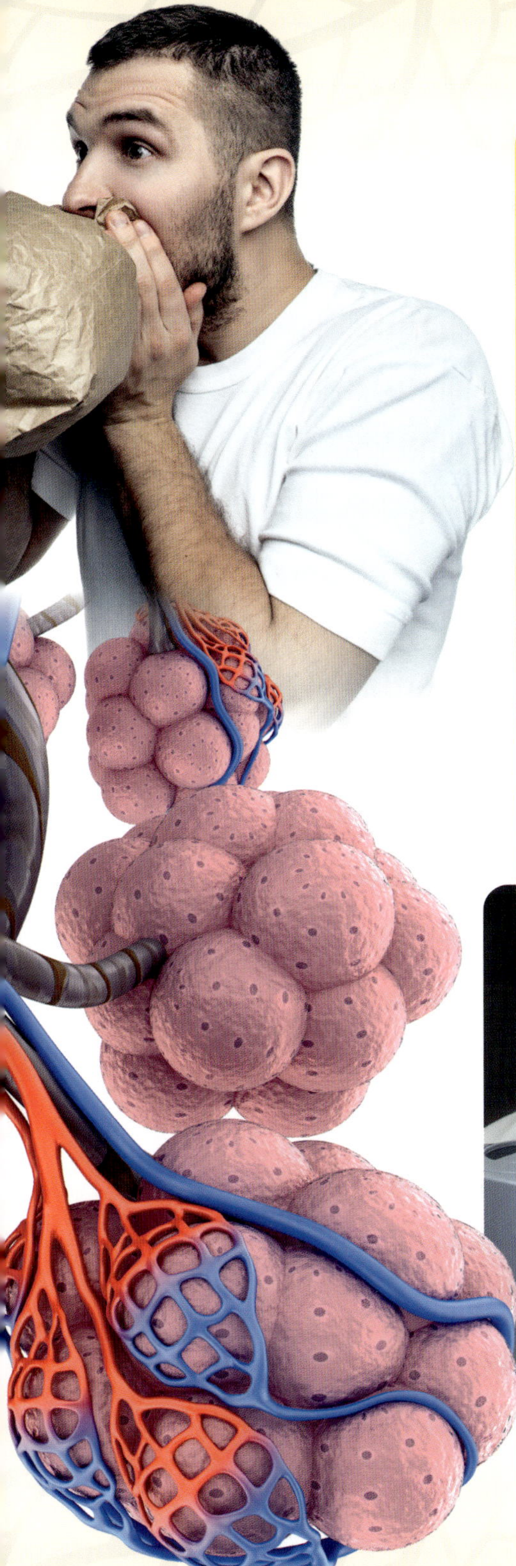

Sleep Apnea

Sleep apnea is a condition where a person does not breathe properly during the night. This may happen because the airway becomes blocked by too much tissue at the back of the throat. It can also happen if the brain's breathing control center does not tell the muscles to breathe.

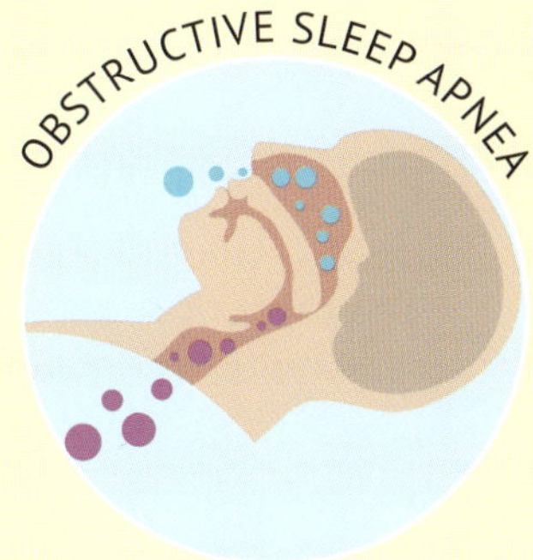

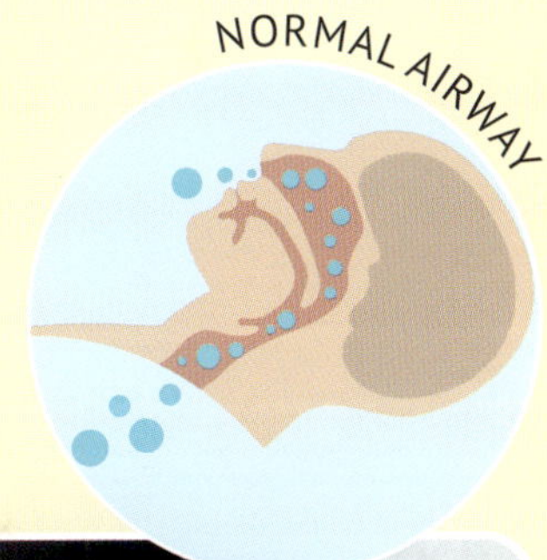

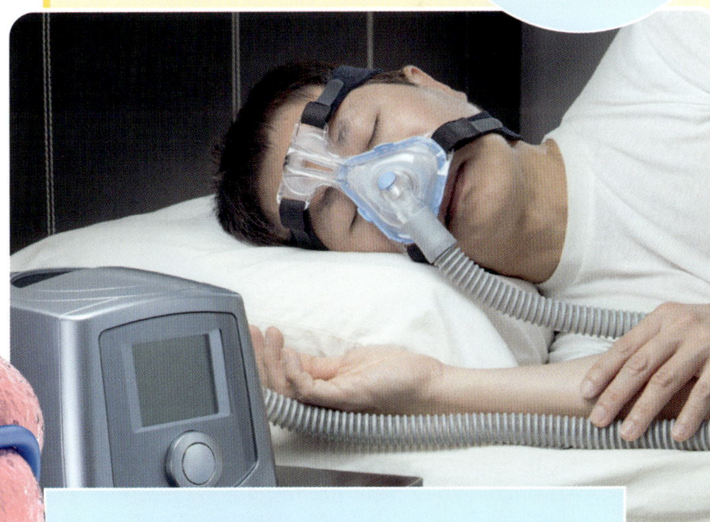

A continuous positive airway pressure (CPAP) machine prevents airways from collapsing. It is used by people who have sleep apnea.

Opera singers do breath exercises and are trained in how to control their breath. They often have strong chest-wall muscles and good lung capacity.

Creating Sound

Without the respiratory system, we would not be able to make sounds. Imagine a world with no talking, humming, yelling, or singing. The air coming in and out of our bodies across special structures in our throats is what makes sound possible.

The larynx is the tube that lets air pass from the pharynx, or the throat, to the trachea on the way to the lungs. The larynx is made of cartilage. Cartilage is a type of firm tissue that is more flexible than bone. The larynx contains two folds of tissue called vocal cords. That is why the larynx is sometimes called the voice box.

As we exhale, air passes over the vocal cords, making them vibrate. The vibrations make sound waves. The muscles move the cartilage around the vocal cords, pushing them together. If you relax your throat and slow the air going over your vocal cords, your voice will sound deeper. Tightening the cords and pushing air across them faster makes a higher-pitched sound.

The shape of each person's face, mouth, larynx, and vocal cords are all unique. That is why people have **distinctive** voices. It is also why people can sing in different ranges. Longer and wider vocal cords produce lower singing voices. Men typically have longer and wider vocal cords and often sing lower notes. Women who sing the highest notes in a song are called sopranos. Sopranos often have shorter and narrower vocal cords.

Some people with vocal cord damage learn sign languages to communicate.

Singers are taught how to properly use their breath in singing lessons.

Stuttering

Repeating certain sounds or having silent blocks is called stuttering. It is caused by the tightening of the vocal cords. This may cause them to "lock," which prevents air from being released and sound being made. People who stutter are often encouraged to try to slow their speech and breathing. This can relax the muscles around the vocal cords and create better airflow.

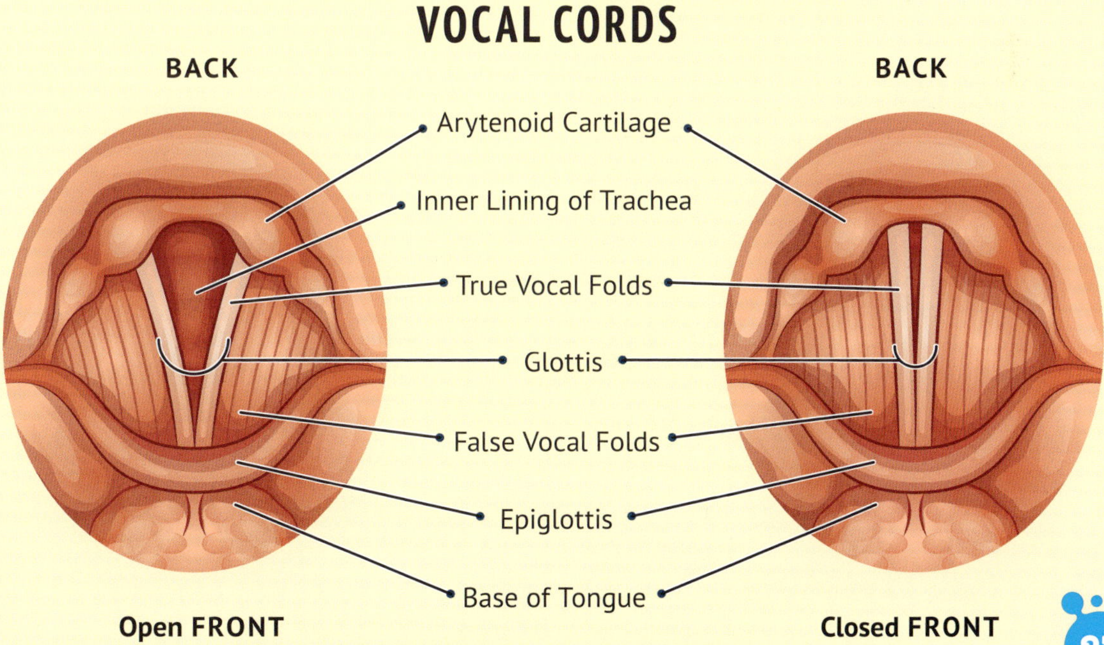

The Nose Knows

You may not think it, but smelling is a function of the respiratory system. This is because we need our noses to smell. The nose, nostrils, and **nasal cavities** are where your body senses and identifies good and bad odors.

Smelling begins when air enters the nose. Up on the roof of your nasal cavity is a small patch of tissue about 1 inch (3 cm) square. It is called the olfactory epithelium. It has special **receptor cells** in it. There are about 6 million odor receptors in your nose! These receptor cells detect odor molecules. Then they send signals along nerve cells to your brain.

The olfactory bulb is where your brain processes smells. It is underneath the front of your brain, just above your nasal cavity. Your brain learns to recognize pleasant smells from unpleasant, and harmless smells from dangerous ones. Identifying smells is how your brain learns about your environment. Your sense of smell can keep you safe. The smell of smoke or rotting food is a warning to avoid those situations.

Smelling is also a big part of tasting. This is because as the tongue is identifying the taste of the food, such as sweet, bitter, or sour, the nose is also identifying the food by its smell. The two work together and combine the information so that flavors can be recognized.

Smell and taste work together.

Dogs have up to 300 million olfactory receptors in their noses. That is why they can sniff out snacks as well as diseases.

OLFACTORY SYSTEM

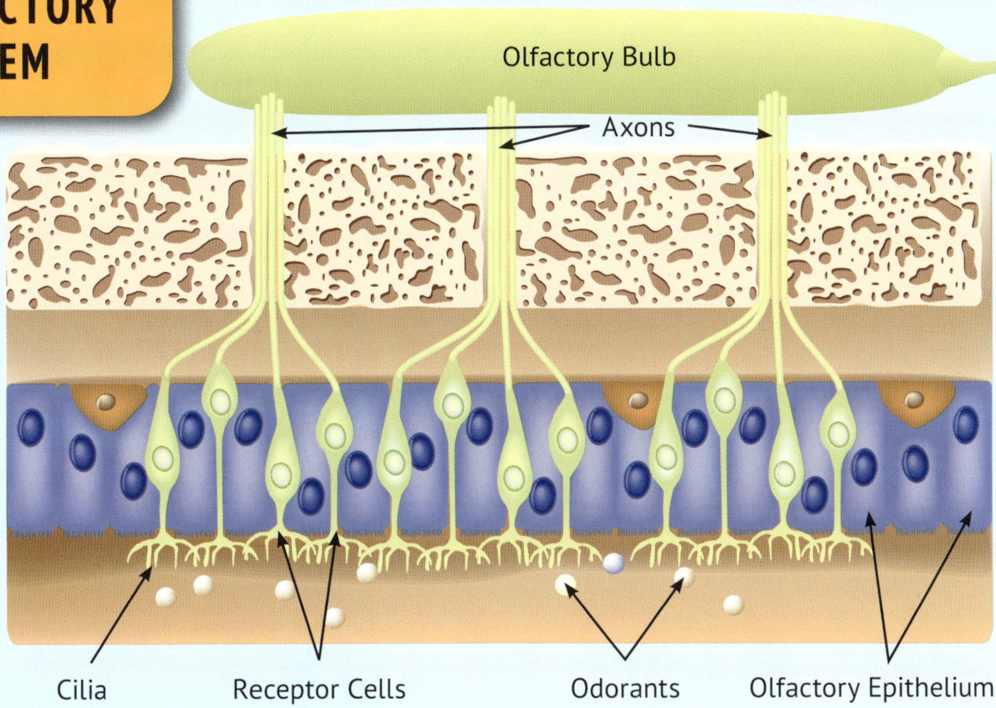

Smelling and Memories

Why do smells evoke stronger memories than other senses? When you see, hear, touch, or taste, those messages go to the brain's relay station, called the thalamus. It then sends the information to the proper areas of the brain for processing, including memory. But smells bypass the thalamus and go directly to the smell center, called the olfactory bulb, which is directly connected to memory.

CHAPTER 3

When It Doesn't Work

Problems with our respiratory system can range from mild to life-threatening. Many different diseases and conditions can affect how well we can get air into our lungs, how well our bodies harvest the oxygen out of it, and how well they expel waste air out again.

Particles in the environment that trigger a reaction in our bodies are called allergens. Allergens can include **pollen**, leaves, food, mold, dust, and microscopic insects. They can cause the lining in your nose to produce a lot of mucus in order to flush the allergens out, giving you a runny nose. Sometimes you start coughing as your system tries to expel the allergens from the trachea or lungs. Most of the time these symptoms go away once the allergen is gone.

Allergens can also produce more severe symptoms. **Inflammation** in the lungs can **restrict** airways. This can lead to wheezing, shortness of breath, and a tightness in the chest. Asthma is the name for the condition when airways swell and narrow. It is usually treated with medications that reduce inflammation and that relax and open airways such as the trachea, bronchi, and the bronchioles.

Allergies can be annoying. Some can be life-threatening when allergens cause reactions that restrict airways.

In rare cases, allergens cause life-threatening problems with the respiratory system. Severe allergies to certain nuts or insect stings can result in swelling that can actually close off airways. This severe reaction is called anaphylaxis. People with this type of dangerous reaction to allergens carry medication that can reverse the swelling quickly.

EpiPens can save lives.

Using an EpiPen for Aniphylaxis

People with anaphylaxis usually carry a device called an EpiPen. It contains a hormone that reduces inflammation. When an anaphylactic reaction occurs, the patient has only minutes to inject themselves with the medication. This is why EpiPens are often carried in a bag or pack on a person's body.

Virus, Bacteria, and Fungi...Oh My!

Infections can affect both the upper and lower respiratory tracts. Viruses and bacteria of upper respiratory infections attack the nasal cavities called **sinuses**, as well as the throat. Lower respiratory infections are found in the bronchi and bronchioles of the lungs.

Upper respiratory infection symptoms include sore throat, runny nose, and cough. The common cold, laryngitis (when you lose your voice), and sinusitis (sinus infections) are some types of these infections. These forms of infections are managed with fluids, bed rest, and over-the-counter medications.

Most lower respiratory infections are usually uncomplicated and handled by the body's own immune system. But when there are complications, they can lead to serious problems. Bronchitis is an inflammation of the lining of the bronchi. Pneumonia is an infection of the alveoli in the lungs. The alveoli fill up with fluid, affecting their ability to transfer oxygen to the blood. The influenza virus, which we often call the flu, can cause both bronchitis and pneumonia. Strong medications help the body fight these infections.

Testing for COVID-19 using nasal swabs

COVID-19 is one of the newest respiratory infections. The virus can affect both the upper and lower respiratory tracts. In the upper respiratory tract it causes a sore throat, sinus infection, and cough. However, this virus can also cause trouble in the lower tract, as well as other parts of the body. It can lead to pneumonia, heart problems, liver problems, and brain damage.

Coughing

Just like sneezing, coughing is a way for the body to get rid of irritants in the body's airways. Coughs are an important **reflex** that pushes air and particles, such as germs, mucus, or dust, out of the lungs and throat. A cough can travel as fast as 50 miles per hour (80 km/h)!

Lung Disease

The term lung disease refers to many different problems with the lungs. Sometimes people are born with lung problems. Other times the environment or a person's lifestyle can lead to these diseases. Difficulty breathing or getting oxygen to the blood can severely restrict daily life.

Cystic fibrosis (CF) is an inherited disorder. That means it is passed down through generations. CF causes thick mucus to build up in the body, including in the lungs. Children born with CF have trouble breathing because the mucus blocks their airways. In severe cases, people with CF sometimes need a **lung transplant**.

Other diseases target the airways, such as chronic obstructive pulmonary disease (COPD). People with COPD have trouble breathing and often cough or wheeze. Lung diseases such as emphysema and chronic bronchitis are types of COPD. With emphysema, the alveoli break down over time. Chronic bronchitis sufferers struggle with inflamed bronchi. These diseases cannot be cured but are managed with medications.

Cancer is a disease that can affect any part of the body. It is the abnormal growth of cells, sometimes into a large mass called a tumor. Lung cancer is the most common cancer worldwide. Since the lung tissue works the same in all parts of the organ, damaged parts can sometimes be removed. People who have some lung tissue removed can still live normal lives. Smoking is one of the biggest causes of lung cancer.

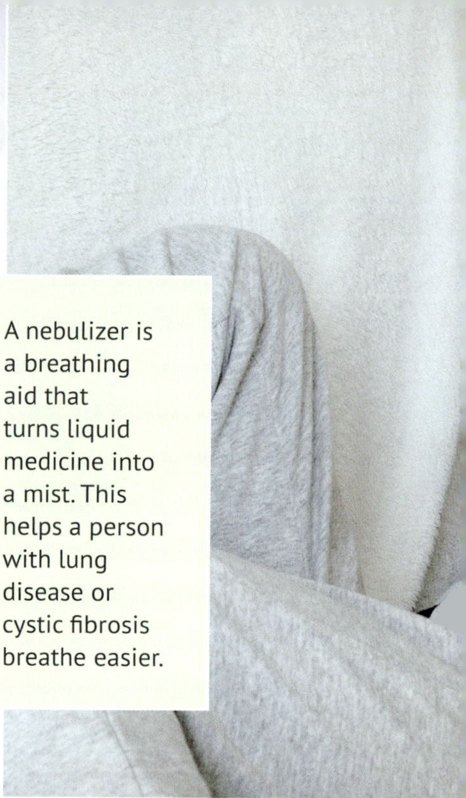

A nebulizer is a breathing aid that turns liquid medicine into a mist. This helps a person with lung disease or cystic fibrosis breathe easier.

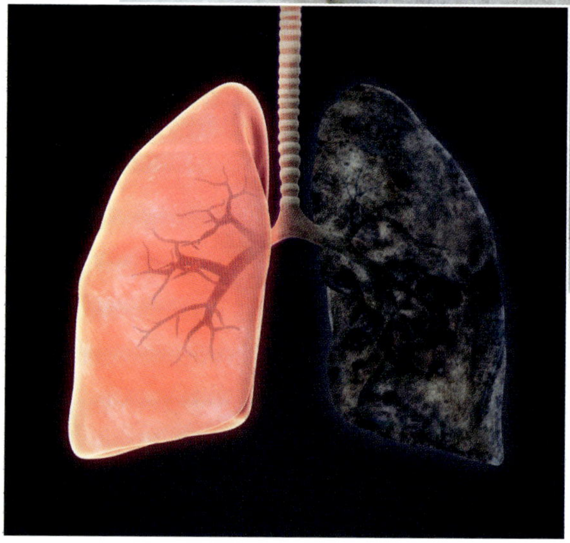

Cigarette smoking causes respiratory diseases that make breathing difficult.

Key Figures

Dr. Dorothy Andersen was an American doctor working in New York. In 1938, she was the first doctor to describe the condition of cystic fibrosis. Before this, it was thought to be a gluten allergy known as celiac disease. Her work led to a simple test to diagnose CF that is still used today.

CYSTIC FIBROSIS

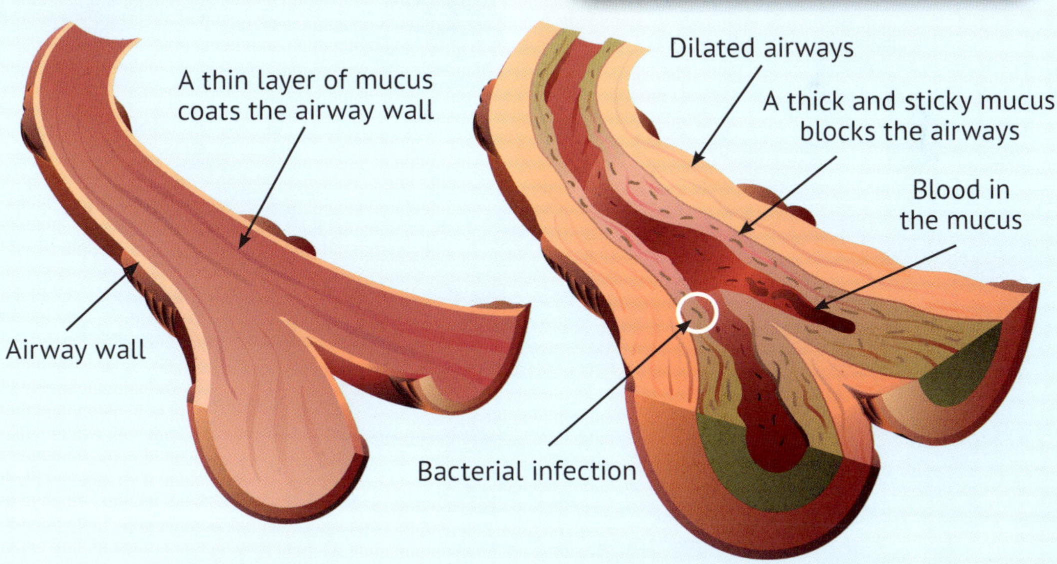

Normal airways **Airways affected by cystic fibrosis**

Lung Damage

It is not just disease or infection that can damage lung tissue. The environment, unhealthy choices, or harmful behaviors can also cause damage. Sometimes our jobs or natural disasters play a part in risks to our respiratory system.

As people age, their bodies naturally begin to weaken and break down. In the lungs, alveoli can lose their shape and become baggy. They don't work as effectively. The diaphragm can also become weaker, making it harder for it to pull air in and push air out of the lungs. These things can make breathing more difficult.

Asbestos is a mineral found in nature that is made up of tiny fibers. If these fibers are breathed in, they can get lodged in the lungs and cause damage. Silica is a very common mineral. When rocks containing silica are cut or ground up, dust is created. Silica dust that gets into the lungs can cause scar tissue, making it difficult for the lungs to work. Smoke from forest fires also contains small particles that are harmful when they enter the lungs.

Asbestos was used in the past in some building materials and construction. It can only be removed by people who are specially trained and who wear protective equipment.

The dangers of smoking have been well-documented in many studies. Tobacco in cigarettes, cigars, and pipes contains thousands of chemicals, including at least 70 that are known to cause cancer. Smoking electronic cigarettes, or e-cigarettes, is called vaping. An e-cigarette is a device that heats liquid into an **aerosol** that users inhale. The liquid contains harmful substances, including certain flavors that have been linked to serious lung disease. It also contains very fine particles that can be inhaled deep into the lungs, causing damage.

Firefighting is a dangerous job both during and after fires. Exposure to harmful chemicals while fighting fires can cause lung damage.

What Is EVALI?

EVALI stands for E-cigarette or Vaping use-Associated Lung Injury. A growing number of severe lung illnesses such as COPD have been linked to vaping.

CHAPTER 4

Keeping the System Healthy

Our bodies need energy, nutrition, and water to move, grow, and heal. Keeping the respiratory system healthy begins with keeping the whole body healthy.

Eating a healthy diet is the best way for the cells and tissue of the respiratory system to get the **nutrients** they need. Lots of different fruits, vegetables, and protein will usually provide all the different vitamins and minerals needed. Studies have shown that vitamin C can help protect against lung disease. Vitamin C is found in citrus fruits like oranges, as well as chili peppers and kale.

Regular exercise is a way to keep muscles strong and help deliver oxygen and nutrients to tissue. The respiratory system especially needs strong muscles as they help the body to inhale and exhale. Exercise also increases blood flow to the lungs and improves lung capacity.

Infections can cause permanent and **irreversible** damage to the respiratory system. Avoiding infections is an important way to maintain the health of the system. Washing our hands to remove germs, handling food safely, getting vaccinated, keeping wounds clean, and even using insect repellents are all ways to prevent infections.

How to Increase Lung Capacity

Certain exercises can improve lung capacity, such as practicing using the diaphragm to let the lungs expand, often called "belly breathing." Pursing the lips to slow down breathing and keeping airways open longer allow the lungs to function better and improve the exchange of oxygen and carbon dioxide.

Olympic swimmers such as Michael Phelps have greater lung capacity because of how they train their lungs and bodies.

Belly breathing is a way to improve the strength of the diaphragm.

Fact Check

Do athletes have bigger lungs?

FACT: Athletes have the same size lungs as everyone else, but their lungs work more efficiently. Their training increases the volume of air that they can inhale with each breath. Their lungs are also more effective at exchanging gases in the alveoli, letting more oxygen reach their muscles to give them more power, strength, and **endurance**.

Butt Out!

When it comes to respiratory health, experts say the most important thing is: Don't smoke! That is because smoking harms nearly every organ in the body, especially the lungs. It also causes many diseases and reduces the overall health of smokers.

The facts on smoking are **grim**. Studies show that smoking causes more deaths each year than illegal drug use, alcohol use, gun incidents, and car accidents combined. Smoking causes about 9 out of 10 lung cancer deaths and 8 out of 10 COPD deaths. Lung cancer isn't the only risk for smokers. Smoking can also cause cancer almost anywhere in the body, including the bladder, blood, liver, and stomach. Even when smoking doesn't cause death, it increases the risk of other dangerous conditions such as heart disease and strokes. If someone has asthma, it can trigger an attack or make an attack worse.

Physical activity, such as running, hiking, walking, or playing sports, helps improve lung function.

Fact Check

Does smoking harm other people?

FACT: Second-hand smoke contains more than 7,000 chemicals. Hundreds of these chemicals are toxic and about 70 can cause cancer. Smoking during pregnancy results in more than 1,000 infant deaths each year.

The good news is that as soon as a person stops smoking, the body begins to heal. Twenty minutes after the last cigarette, heart rate and blood pressure return to normal. Within eight hours, the body's oxygen levels increase and lung function begins to improve. As the lungs start to heal, the ex-smoker will likely breathe easier, cough less, and have more **stamina** when they exercise. Smelling and tasting improve because smoking deadens nerve endings in the mouth and nose. Quitting improves blood flow around the body so skin looks healthier. Ten years after quitting, the risk of developing lung cancer is cut in half. The risk of other cancers decreases too.

Quitting smoking helps improve blood pressure and general health.

Finding Answers

Treating lung diseases depends on early **detection** and treatment. Doctors use a variety of equipment and tests to identify problems. Getting a look inside the lungs allows doctors to see if a tumor is forming, if airways are tightening, or if there is damage.

X-rays are a type of high-energy **radiation** that can pass through most objects, including the body. As the light hits **dense** tissue or bone, it creates an image. Chest X-rays can show cancer, infection, or if air is collecting in the space around a lung, which can cause the lung to collapse. Computed tomography (CT) scans use low-dose X-rays. With this type of scan, the X-ray beam moves in a circle, allowing the radiologist to look at different "slices" of the lungs.

An MRI is a magnetic resonance image. This technology uses a magnetic field and radio waves generated by a computer to create a very detailed image of organs and tissue in the body. It is not as good for examining lungs as a CT scan, but it is useful for people who may have had too much radiation from repeated X-rays.

Pulmonary function tests examine how well the lungs are working. These tests measure several different levels, including:

- How much air is moved during normal breathing
- How much air is moved per minute
- How much air is left in the lungs after exhaling
- The total volume of air the lungs can hold.

These numbers can show if there might be something blocking airways or whether or not the lung tissue or chest muscles can expand and contract properly.

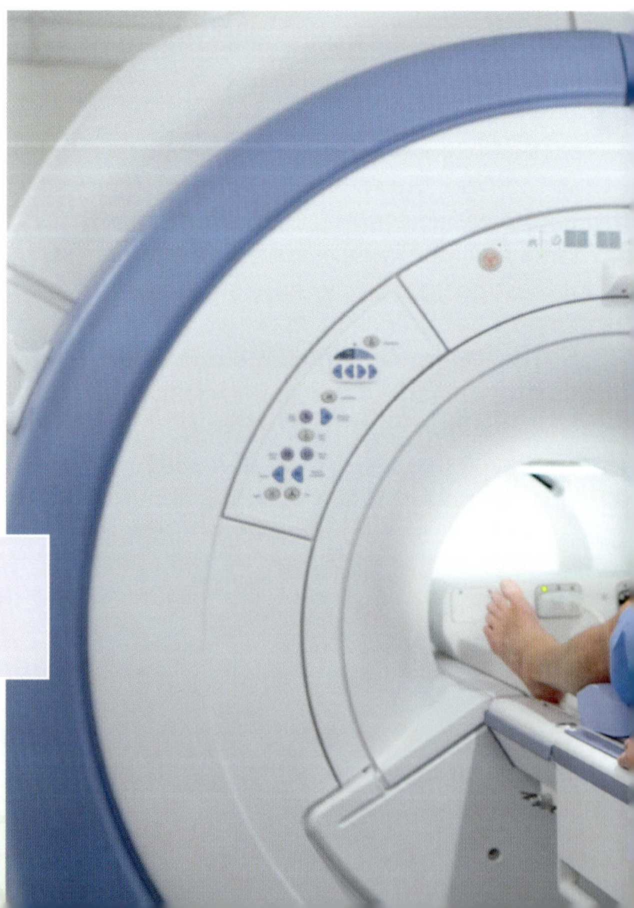

MRIs are loud, enclosed machines that take images of the body's inner tissue.

Sputum

Sputum is a term for the mixture of saliva and mucus coughed up from the respiratory tract. By studying sputum, doctors can identify viruses, bacteria, and fungi that might be in the respiratory system.

This sputum smear magnified under an LED microscope shows tuberculosis (TB) bacteria. TB is an ancient disease that affects the lungs and other parts of the body. It can be deadly, especially if not treated. In 2020, 10 million people in the world developed TB and and an estimated 1.5 million died from infections.

CHAPTER 5

Looking to the Future

Keeping us breathing efficiently and identifying breathing risks are some of the goals of new technology. Smart devices are being developed to monitor for dangerous conditions or the beginning of breathing problems.

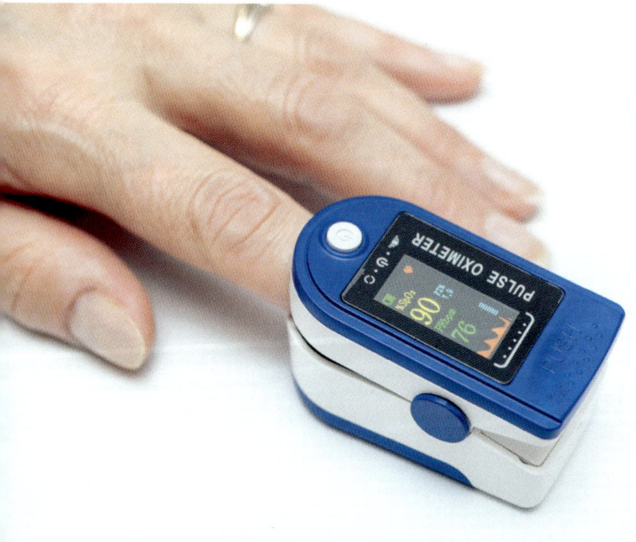

Pulse oximeters are so efficient now that people can buy them at pharmacies and use them at home. They measure the oxygen saturation level in a person's body without having to use needles or take blood samples.

Researchers in Madrid, Spain, have designed a smart shirt that senses movement in your chest and abdomen. It can detect slight changes that might signal breathing distress. The information can then be sent to a patient's phone or a doctor's computer. This can warn patients of an oncoming asthma attack. It can also gather information that allows treatment to be customized to an individual's symptoms or inform doctors when a patient's condition is getting worse.

Another type of "wearable technology" is an electronic sticker. This skin-like sticker can monitor a person's respiration, muscle activity, and heart rate all at the same time. The sticker is flexible, so it stretches with the skin and does not restrict movement. It sends data wirelessly to a receiver clipped on a person's clothing. Some engineers are even trying to embed sensors in tattoo ink!

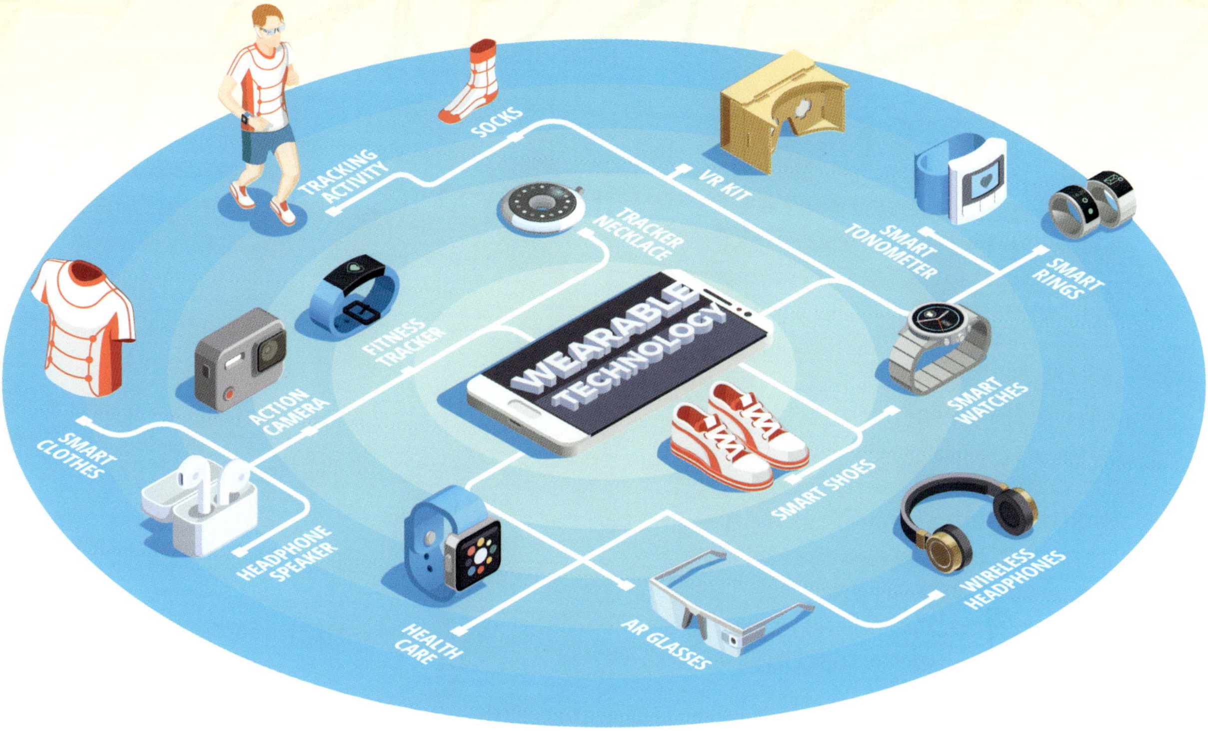

Researchers are looking into the effects of breathing wildfire smoke. The chemicals and microparticles that are inhaled travel through the bloodstream and reach the brain. This can cause many health risks, from premature aging to forms of **dementia** and depression. Firefighters and people living in frequent bush fire areas can buy small air-quality sensors. These sensors are the size of a coffee cup and have **lasers** that count particles in the air. They can warn of dangerous breathing conditions.

Mice of Hope

To date, people suffering from a disease called pulmonary fibrosis have had no cure. However, studies on mice show that a new medicine has reversed pulmonary fibrosis. Researchers are very hopeful that this may lead to a cure for humans.

No Words

Some people cannot speak because they were born with vocal cords that did not form properly. Others lose their voice box to cancer or injury. Scientists have now developed a thin, artificial "voice box." This device can be attached to the neck like a temporary tattoo using just water. It can convert the slightest movements into sounds. People currently testing the device can make simple words such as "OK" or "No." Researchers are working to make this device more advanced to help the wearer regain normal speech.

Scientists at McGill University in Montreal, Canada, have developed a new way to help repair muscle tissue such as vocal cords. It uses biomaterial. This is a human-made substance that **interacts** with living tissue. McGill scientists can inject this biomaterial into the body where it helps heal injured organs, including vocal cords.

A loss of the sense of smell can affect a person's enjoyment in life, such as the aroma of a favorite meal or the smell of flowers. It can also increase the risk of injury or disease when people cannot detect food that has gone bad or a gas leak. Researchers in Virginia are working on an **implant** that could do the work of the olfactory bulb. It would take in sensory data and tell the brain what the smell is.

VOCAL CORD DISORDERS

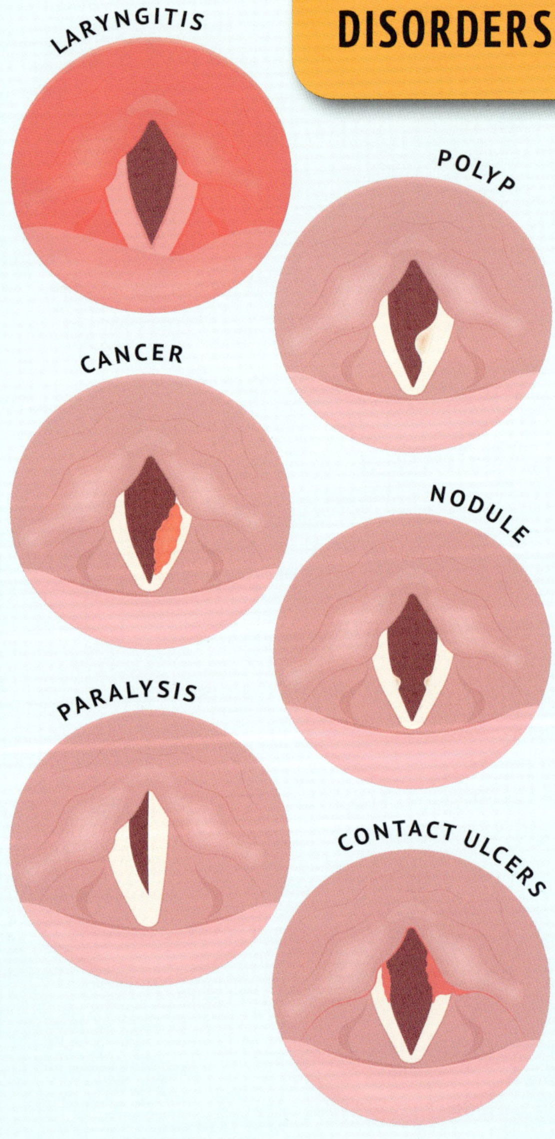

The voice box contains the vocal cords. The cords vibrate and air passes through them from the lungs to produce the sound of a person's voice. When they are injured, or even inflamed by laryngitis from a cold, the sound of a voice can change. When both vocal cords are injured, a person may have trouble breathing.

Instant Cure for Hiccups

Hiccups can be annoying and there are all kinds of tips for how to cure them: scare someone, drink from the wrong side of a cup, or hold your nose while drinking water. Scientists have now created a straw that instantly cures hiccups. The device has a pressure valve that requires the diaphragm to contract more to sip water, which stops the **spasms** that cause hiccups.

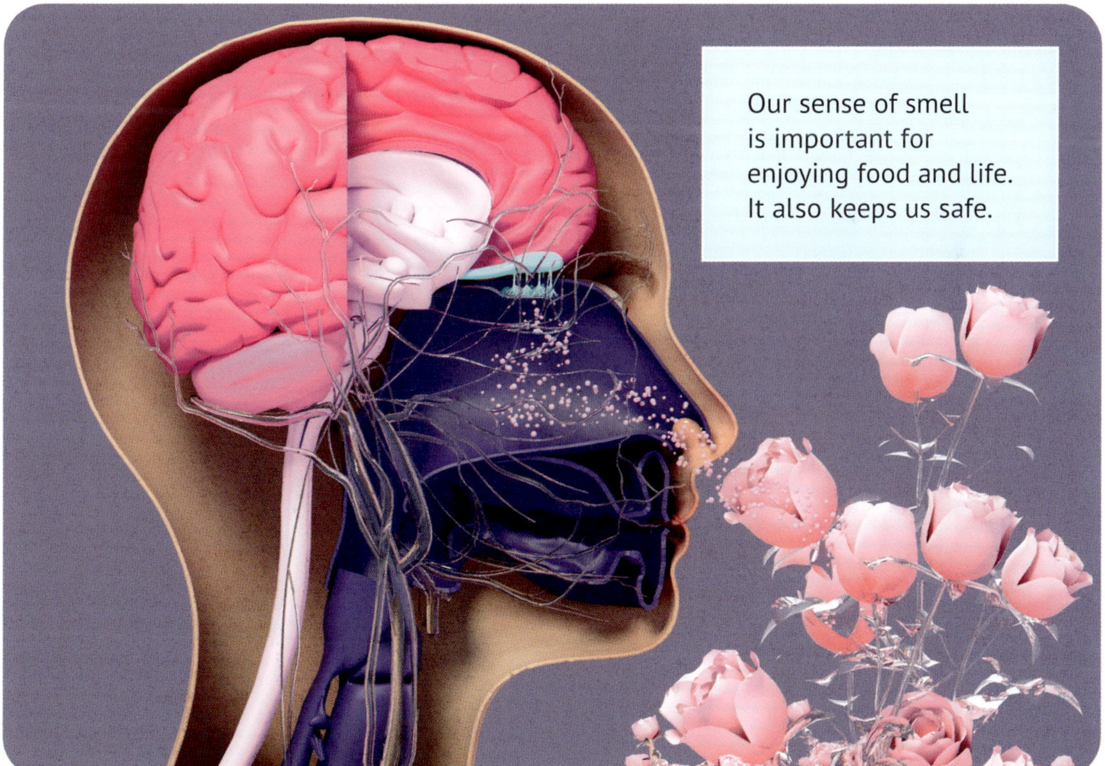

Our sense of smell is important for enjoying food and life. It also keeps us safe.

Take a Deep Breath

Chronic respiratory diseases such as COPD and asthma affect more than 500 million people worldwide. Treatment for these conditions often includes using an inhaler. This device delivers medication to the lungs directly, instead of going through the bloodstream in a pill.

For inhalers to work effectively, the dose of medication has to be carefully measured and given on a strict schedule. People with busy lives, seniors with memory problems, or children who are distracted can sometimes forget to take their medication on time or in the right amount. Smart inhalers can be set up to remind users of their next dose. When smart inhalers are connected to phones, they even alert users if they have left them behind when going out. They can also gather all kinds of data, or information, from use. This includes how often doses were late, the activity level at the time of use, and even the user's location. This data can help doctors adjust a patient's plan to make it more effective for each individual.

Listening to the lungs often gives doctors a good idea of a problem. In normal lungs, a doctor can hear a soft blowing in the lung tissue or air moving through a pipe in the trachea. Abnormal sounds are wheezing or crackling in the lungs. These are signs of restricted airways or fluid in the lungs. Older-model stethoscopes take in all sound equally. It can be difficult for doctors to hear over a noisy emergency room or a crying patient. New smart stethoscopes have five programmable microphones inside the head of the stethoscope that can focus in on the chest and not the room. A chip inside the stethoscope can strip out room noise so the doctor can get a clear idea of what the lungs and breathing sound like.

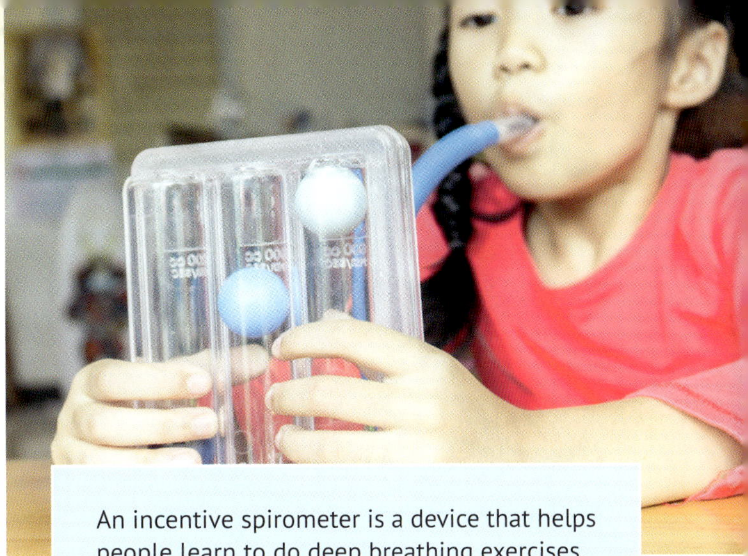

An incentive spirometer is a device that helps people learn to do deep breathing exercises.

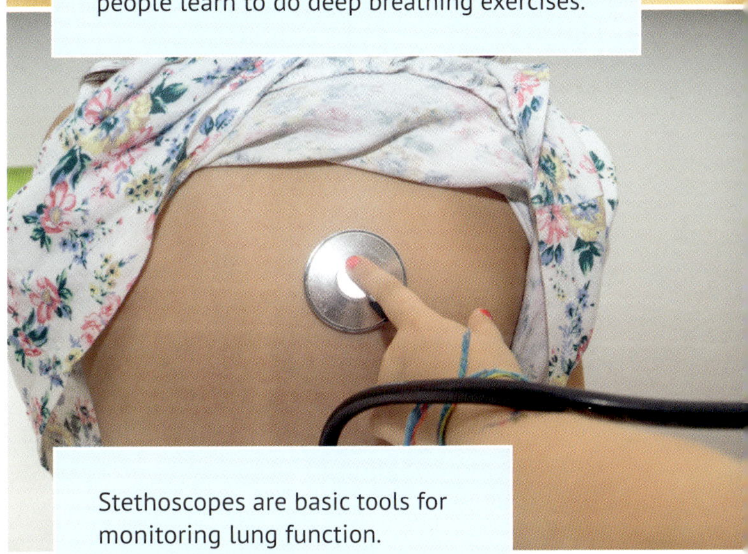

Stethoscopes are basic tools for monitoring lung function.

Phone and watch apps can help people check their heart and lungs.

Glossary

absorbed Taken in or soaked up

aerosol A fine spray in the air

allergens Substances that cause a reaction in the body

amphibians Animals that can live both in water and on land, such as frogs

autopsies Examinations of bodies to figure out the cause of death

brain stem The central trunk of the brain that attaches to the spinal cord

chemical process A reaction that takes place inside living things

contract To become shorter and tighter

controversial A topic people disagree about

deflate To grow smaller and flatter

deliberately On purpose

dementia A brain disorder that causes memory loss

dense Packed close together

detection Identifying the presence of something hidden

diaphragm The dome-shaped muscle under the lungs

distinctive Having traits that make it easy to identify

endurance The ability to do something difficult for a long time

esophagus The tube that connects the throat to the stomach

expel To force something out

grim Something that is worrying or without much hope

hormones Chemical substances that deliver messages around the body

humidity The amount of water droplets in the air

implant A human-made device or tissue used for repairing or replacing part of the body

inflammation A condition in which part of the body becomes swollen, red, and painful

interacts Acts together with something else

irreversible Cannot be undone

irritant Something that causes discomfort

lasers Devices that create intense beams of light

lung transplant Replacing a damaged lung with a healthy one from a deceased donor

microbes Tiny living things

nasal cavity The space inside the nose

nutrients Substances that provide vitamins and minerals that the body needs to be healthy

particles Very small pieces of something

polio A life-threatening virus that causes parts of the body to be unable to move

pollen Fine dust that is given off by flowers and that is needed for plants to reproduce

radiation Energy given off as waves or particles

receptor cells Cells that take in information

reflex An action that is done without thinking

restrict To limit space or movement

Roman Empire A powerful ancient civilization that ruled much of Europe for almost 1,000 years

sinuses Eight spaces connected to the nasal passages that make mucus

spasms Sudden, uncontrollable movements

stamina The ability to do something for a long time

stethoscope An instrument used for listening to someone's heart or lungs

Learning More

BOOKS

Jango-Cohen, Judith. *Your Respiratory System.* Lerner Classroom, 2012.

Mason, Paul. *Your Breathtaking Lungs and Rocking Respiratory System.* Crabtree Publishing, 2016.

Simon, Seymour. *Lungs: All about Our Respiratory System and More!* HarperCollins, 2007.

WEBSITES

Have fun with Bioman's interactive activity. This respiratory journey takes oxygen from the air all the way through the respiratory system to body cells and then helps carbon dioxide find its way back out: **https://biomanbio.com/HTML5GamesandLabs/Physiogames/respiratory_journeyhtml5page.html**

Take a quiz with KidsHealth to see how much you have learned about the respiratory system: **https://kidshealth.org/en/kids/rsquiz.html**

See if you can label the entire respiratory system and do other fun activities with Learn-Biology: **https://learn-biology.com/respiratory-system/**

Index

About the Author

Natalie Hyde has written more than 100 fiction and nonfiction books for kids. She has always been fascinated by the extraordinary machine we call the human body. Natalie lives with her family, their dog Louie, and a flock of Rhode Island Red chickens called "The Ladies."